Primer to Aleodeology

©Mark Cater 2016

<markcater@gmx.com>

FORWARD:

This book is the result of several years of effort and planning which began almost 24 years ago.

A reader may find the concepts disturbing at first, as they should be. This is because the information is in conjunction with my other book CONCEPTS BEHIND ALEODEOLOGY. You do not have to read this book in order to comprehend the science behind the Primer to Aleodeology. All you require is an Open Mind.

I find that if a serious reader of this book uses commons sense you will see the truth of it`s logic right away as being fact. The information may, however seem complicated –but rereading and using logical deductions will show you that the logic involved is sound. You may find however, a typo or two. That`s Ok, not everything is perfect, so keep that in mind.

I spent a lot of time verifying the information here, in this book myself and I believe it is sound.

Now you will find also that naysayers will appear out of the woodwork –these are the disinformation specialists; of course once they too find that what is presented here is in fact

> real, it will more than likely make them reconsider that they were too judgmental without looking into the matter.

So, I will leave the rest up to you – I present the truth about how Gravity is identified as and what it really is as opposed to what conventional science has been led to believe.

I think any reader of this work will start asking serious questions about why scientists have not laid hold on what I have discovered, in many ways the math they use is self-referential. That makes it difficult to identify; especially when it is value-based and manually based. Terrence McKenna gave a lecture on the Nature of Magic, what is it? Very logically he explained that the world (and I assume that this extends to the universe in general) is, in fact language, because what we get back each time are words. Perhaps the language has yet to be fully be made known but if we delimit ourselves to explaining everything physical and metaphysical to math, we doom ourselves to a plethora of ridiculous aspects through which new discoveries will hardly every become reality. Open the doors open up for new information.

Table of Contents

I. Introduction
II. How Many Die Rolls Until You Get A Repeat?
III. Repetitious Input
IV. Gravity Identified
V. A Simple Lesson From Scripture
VI. Energy is Manual, Potentiality is Automatic
VII. Mass Causes Degrees of Freedom through Potentiality
VIII. Recent Findings
IX. Still More to Learn

INTRODUCTION

The Idea has come up of lately that space has time only with respect to energy. Energy as it stands is manually based and it`s measurements rely primarily upon values, which differs extremely from symbols based with respect to the true source of information itself. This form of information encompasses massive amounts of data that cannot be handled with conventional means and differs from knowledge.

What is being measured is velocity/speed with respect to space over a predetermined distance in an amount of time over this distance; as the manual push-action propagates the mass or system of masses **THROUGH** space, interaction with space is negligible. That is all that is occurring. Time then is more a function of energy with respect to space rather than a property of space, although space accommodates energy to operate through it timitically.

Force as depicted by Newton, is merely energy transfer from a manually based source and therefore the source is the force. This source cannot compensate for potential energy, which automatically becomes established with the establishment of manual energy input producing a repetitious output to maintain a velocity or a change in velocity.

By repetitious input producing a repetitious output it is meant that repeated input is required to maintain velocity or change in velocity in order to avoid the automatic establishment of potentiality against the manual establishment of energy.

With respect to the mass, potential energy impedes the mass from self acceleration, thereby pre-requiring the input of energy transfer from the manual source to be repetitious – otherwise absent of this repetitious input potential energy would result bringing the mass back to rest.

Irrespective of whether the mass experiences uniform velocity/speed or acceleration manual input of energy to the mass must always be repetitious to avoid the feedback of potential energy when manual energy is applied to the system, the only difference being with acceleration is that of more "power" being pre-required to be put in to the system to perpetuate to output of the system incrementally.

Thus manual input of energy to a system to produce the output through the manual input seen by an external observer is plagued by the automatic result of feedback of potential energy from the system, necessitating repetitious input of energy to evade feedback from the system`s energy input.

"Feedback" as it is seen is related to the mass itself with respect to gravity in that it can be deemed as gravity, for there is no identification as to where it originates. We can comprehend this when we imagine a system at rest and it's buoyancy in outer space, apart from the rotations of the planet and sun. So we can estimate to a small degree that the rotation of Earth and the sun are at least "partly" responsible for this, while in outer space the system is detached from the "pull" associated with the larger mass's elliptical revolution as well as the sun's own pull upon Earth, leaving the system to float or be constantly pulsed forward, producing slower than light velocities but eventual breakdown of the system.

All measurements are value-based; originating from energy manually established observations. Because these value based measurements of energy are manual, they are therefore self-referential and are not dependable for realizing highly advanced technologies that can be accomplished. Thus there is no real "interaction" with space; there is only timitic propagation of the energy or the system propagated by manual energy THROUGH space. Thereby space is not a function of time; rather time is a function of energy with respect to it as it is manually established.

If the world-line (a path in Einstein space-time (a term used by Government hired scientists in a variety of ways to "unify" two components into one force or concept representing very different phenomena, similar to unifying the words Grandma with a Cyborg into Grand-Borg., i.e. energy and mass can be unified into one word as energy-mass or mass-energy to describe "their" interpretation of energy transfer without looking at the two phenomena as interdependent, (much of which is false and should be rejected regardless of scientific seniority or credentials. For example, Gravitons; another false word-unification of particle and gravitation)) of a moving particle is parameterized thus:

$$(t,x(t),y(t),z(t))$$

We can obtain the "4-velocity" of the particle:

$$(1,dx/dt,dy/dt,dz/dt)$$

The t-axis or time-axis velocity component is 1, a dimensionless number. Now there are relativists who will insist that it is perfectly acceptable to express velocity in time with a dimensionless number but the rest of us with our head on our shoulders, know that it is not true. We know that a dimensionless number such as 1 has absolutely no meaning in as far as expressing velocity or any sort of change. Velocity must be given in units of velocity such as meters per second or whatever standard of units a user audience handles. For this reason, there is no motion in so-called unified Einsteinian Space-Time.

[There are people who will insist that there is nothing wrong with saying that motion in time occurs at the rate of 1 second per second. First of all, dt/dt does not equal 1 second per second. The units cancel out. Second, dt/dt is always the same (1) regardless of the actual rate of velocity.]

Note that I put 4-velocity in quotes above. This is because it is not a velocity at all since nothing can move in space-time. There is only 3-velocity in 3-D space: (dx/dt,dy/dt,dz/dt), t being a mere evolution parameter. True 4-velocity (dw/dt,dx/dt,dy/dt,dz/dt) would require a 4-D manifold having 4 *spatial* dimensions and no time dimension. Now that is an interesting idea, four spatial dimensions, an idea I certainly would not object to.

But time-travel? Many suggest Absolutely not!

One of the amazing things about this time travel business is that a position in Einstein space-time is usually represented by (ct,x,y,z). What this means, is that every second a body moves exactly 299792458 meters, or a light-second in the fourth dimension. (This tool is convenient in explaining what is called a particle's light cone because it allows the sides of the light cone to slope at 45 degrees.) However, note that, using this convention, the fourth dimension is no longer a temporal dimension but a spatial one. Why? Because ct resolves to meters, not seconds. Does this means that time travel is suddenly allowed? Of course not since c is a constant and t is not a variable. It is just that most relativists cannot bring themselves to the point of accepting a fourth spatial dimension. They are forever stuck with space-time for historical reasons.

There is an unyielding mental barrier that I am still in the process of identifying. I wonder if it is just intellectual inertia or a vestige of the historical origin of relativity, kind of like the way an atom is not really an atom in the literal sense of the word. Somehow, I don't think so. By acknowledging the unchanging nature of Einstein space-time, the vast majority of relativists would have to admit that they have been teaching crackpot science (the teaching of geometry as an explanation of gravity) from the beginning. That is unacceptable, of course. Still, it is no excuse to conjure up all sorts of witchdoctor nonsense and retard progress in our understanding of gravity for close to a century without conclusive evidence that it is a force and yet somehow as to be unidentified as to what it actually originates as.

Note: the representation of a position in Einstein space-time is conventionally given by (ict, x, y, z) where i is the square root of -1, an imaginary number (crackpottery never ending). However, this is a mere detail, one which does not take away from the changeless nature of Einstein space-time.

In addition to this, feedback THROUGH space disables simultaneity timitically as it is THE cause for degrees of freedom (abbreviated *df*). For example, take a metal rod representing a line-segment representing a section of space.

If we were to enable the rod to self-rotate through a 360 degree circle and continue to rotate the rod through every circle, it could possibly occupy (equidistant from it`s center), in one direction, it could then occupy all possible outcomes; forming a sphere - with the exception of the opposing direction. This would tend to indicate that the rod would have to come to a complete HALT to begin the identical procedure.

With this single direction, occupancy of all possible outcomes, the sphere formed by the rapid rotations revolving through every circle will pass through all combinations and permutations. Once such is done all that can be done from here are repeats, delimited by feedback. (Feedback here indicates **Pe**).

These repetitions are analogously similar to the repetitions of dice, only that the number of combinations and permutations until repetition transpires is much, much greater in number.

Remember, this is equidistant from the center – any repeat can be induced at any point through the combinations and permutations, so too analogous to dice, all numbers repeat. And similarly like dice they can are knowable and predictable via the repeats themselves giving a 100% expectation. This would surely aide in solvent of negative expectations related to table games; as such let`s explain how this works with die themselves before continuing:

II. How Many Die Rolls Until You Get A Repeat?

Dice tumble, through human factors, they do not roll, coined as such;

What would be the expectation derived from the number of rolls (come-outs) until a standard casino die establishes a number having previously rolled? For instance, assume you roll 1, then 2, and then 1 again. The die repeated the/a number 1 on the 3rd roll. Or if toss 3, 2, 5, 2, then the resulting number 2 shows again, it took 4 rolls to see a repeat.

What then is the average number of rolls until a die repeats a number?

Solution:

The 1st established come-out roll is a unique number and never a (emphasis on "a" as opposed to the) repeat, while the 7th roll always results in a repeated number. So the answer may be estimated to be somewhere in the middle, like **3.5**. This is approximate! The exact answer is **1223/324**, which, is about **3.77**.

We can resolve the problem by working backward. When one has seen all six numbers appear, then we are guaranteed a repeat number in the **1** more roll.

What then if we have seen **5** numbers? There is a **5/6** chance that a duplicate will next appear–in which case we are done in that **1** roll. And there is a **1/6** chance that a new number will appear–in which case we arrive at having seen **6** numbers, plus we add the roll that got us there. In other words, if $E(x)$ is the expected number of rolls after seeing x numbers, we have the following formula:

$E(5) = (5/6)(1) + (1/6)(1 + E(6))$
$E(5) = 1 + (1/6)E(6)$
$E(5) = 1 + (1/6)1 = 7/6$

What if we have already seen **4** numbers? There is a **4/6** chance we will see a duplicate–in which case we are done in that **1** roll. And there is a **2/6** chance we will see a new number–in which case we get to the case of having seen **5** numbers, plus we add the roll that got us there.

$$E(4) = (4/6)(1) + (2/6)(1 + E(5))$$
$$E(4) = 1 + (2/6)E(5)$$
$$E(4) = 1 + (1/6)(7/6) = 25/18$$

Here is a pattern to the calculations. If there have been *x* numbers, then there is an *x*/6 chance a repeat will result—and we are done in that **1** roll. Alternatively there is a **(1 − *x*)/6** chance we see a new number—in which case we get to the case of having seen *x* **+ 1** numbers, plus we add the roll that got us there. Thus:

$$E(x) = (x/6)(1) + ((1 − x)/6)(1 + E(x + 1))$$
$$E(x) = 1 + ((1 − x)/6)E(x + 1)$$

Calculating, we can find **E(3) = 61/36, E(2) = 115/54, E(1) = 899/324,** and finally **E(0) = 1223/324**.

So, when we start out—and no numbers have yet been seen—the expected number of rolls until we see a duplicate is **E(0) = 1223/324**, about **3.77**.

Using the recursive approach further can, of course aide in the potential of wins over losses in abating a negative expectation in games of chance provided we first understand that all numbers of the die are repetitious and that is the basis of so-called probability.

III.   Repetitious Input

Repeats therefore are seen not only in **repetitious input** caused by feedback, Potentiality (i.e. Potential energy *Pe* so-called) but also in the instance of the example of the rapid spinning rod taking up residence in all circles to form a sphere, in one direction, equidistant from the center. Simultaneity therefore **cannot** be accomplished in *both directions* because of feedback, and to add, no matter how the rod itself is re-located, the rod reestablishes similar rotation, again equidistant from the center, the presence of feedback becomes immediately evident automatically to cause these *df* regardless, both in *repetitious input (*see below for a full explanation) and the sphere.

In assuming that feedback is closely related to gravity with respect to the system, indication has it that if gravity were assumed as being automatic, then it cannot be identified within mathematics yet, because the measurements are **self-referential** and value-based originating from **manual energy observations** derived from an initial source, which is also based manually forming the manual input to the system that results in repetitious input caused by the automatic feedback (Pe) in energy transfer establishing displacement. I know this is a mouthful but read it again.

This "repetition" extends to the example of the sphere equidistant from the center- manifest through the rod. We can say that within our three-dimensional space (time as a function of energy) that feedback as gravity is the cause behind **degrees of freedom** where-in **everything** repeats at some point. This outlays any mass occupying space (distance is time) from the movements of a biological organism, right down to the total combinations and permutations of its eye movements a cedar block to larger masses as those of planets and galaxies etcetera, etcetera. There are just so many a mass can go through and this is due to potentiality.

How chemicals, atoms, and quantum energies interact.

Everything repeats.

Because the automatic establishment of potentiality established through manual energy transfer was the oversight, and because it is so negligible it went unnoticed. It would be impossible then to isolate potential energy from its interdependency with energy; they cannot be undermined and the fact is that they are expressions of each other, interdependent. Especially absent of time travel with respect to that barrier caused by potentiality as it applies to the mass of energy rapidly traversing THROUGH outer space.

From this it is clear that because of the instantaneous production of energy's feedback, potentiality, with the establishment of energy to a system, that **self-referential measurements** will **not** identify inexactly what gravity is whether or not; because it is identified as being potential energy or associated with energy as it corresponds to mass powered with manual input that results in repetitious energy input to output.

Assuming this, it can possibly be understood only through an automatic system or automated symbolism disassociated from self referential math based on manual energy observations – in the main perhaps absent of potentiality. In addition there is another other difficult problem in relation to this, that manual input of energy or energy itself cannot be automated in light of feedback resulting from energy transfer.

In more direct terms, an infinite system of energies similar to the infinity mirror would have to be devised to resolve such a problem -- even if the intent were to produce faster than light travel. This is because manual input against potentiality cannot overcome itself. If the manual input that turns to repetitious input due feedback were automatic, it **might** work, but this too would require **the infinity mirror of infinite energies**, each backing the other up from the first.

A few considerations in this is that of somehow causing energy or to the system powered with energy to enable time travitics, if it could be done, it would then be entirely practical to cause such automation because the energy could leave and arrive in a different space at the same time (which is also based to self referential values i.e., (An imaginary reference point indicative of position to mark of the arrivals and leavings about a rotating circumference)) as opposed to arriving and leaving "at" the same position at the same time.

So nothing but automated energy or some other unknown force (not tachyons or gravitons- they do not exist) can abate the manual input resulting in repetitious feedback that might cause the phenomena of advanced technologies capable of enabling superluminal transport, teleportation and the like, or how it can come about THROUGH an automated/automatic symbolism/system.

Information is not four dimensional only, bound to probabilities concerned with the delimitations of Einstein space-time conforming to sets (information theory –Shannon 1948).

If it were, then all information with respect to 4-d/D space and higher echelons of space would of course be manually based and thereby bound exclusively to energy. Since this is understood, it too can only be grasped with respect to its origin (where it originates as opposed to knowledge) via automation.

Throughout my research the convictions I have come across have not lined-up with any known so-called "force" apart from energy, manually driven. Although gravitomagnetism itself possesses automated fields they hold extraordinary promise. However some insight has come via Thoth`s Emerald tablets (although no verifiable evidence as to the actual existence of these tablets have come to light) The writer indicates that space is "conscious" or has a consciousness probably associated with intelligence or an intelligence which differs from our perceptions of what consciousness indicates (the common definitions via mathematical scrutiny). Perhaps, in the writer`s awareness consciousness includes a time-space automatic in the conscious phenomena. Or to further this concept, we do not create it, it is already there just as Iceland is there; we don`t see it, but until we go there.

The writer also seems to signify that space is "divided-by" time (which makes sense) as measurements of time by absolutists of mathematical infallibility are substantiated according to how energy moves THROUGH space 1 dimensionally. (Manually based, pushing action established from an originating source, say the touch of a human finger displacing a small cedar block across the smooth surface of a table). That being said, **no interaction with space is happening apart from the calculus of displacement THROUGH the accommodating space**.

Logic Note: {*Assuming the aspect and possibility of alien tech; (THE) transport method to cover the vast distance of space could **in no practical sense be realized using any manual form of energy**. This tends to suggest the usage of a fifth or higher dimension access to link the vast distance to a conformable and acceptably easy transportation bridge. This further indicates that with such awareness that the aliens either originate from one (or many aliens originate from them) of these dimension or have direct access to and from them interstellar speaking*}.

IV.   Gravity Identified

The so-called scientific definition of Potential-(ity) is spacial (of or pertaining to space), NOT a function of the displacement of mass from the manual input of energy that produces that output. This is because the resultant output of velocity or change-in velocity (more power) is timitically self-regulating by potential to the manual input in the form of so-called gravity (attractive force/pull) executing through the potentiality. (The false term of potential associated as energy, i.e. "potential energy") Potentiality does NOT possess properties of "energy" **until *actualized*** in opposition to manual input of energy to a mass, and therefore with more manual input of power to the mass. The timitic feedback of potentiality requires repetitious input to evade the drag of potentiality, increasing exponentially with exponential input of energy to the mass. Because it is self-regulating, it would be a function of space and NOT a function of any energy establishment, but is coupled to it through manual input making "Feedback" automatic.

* "*i*" is Input of E to mass which is any manual energy.
 "*o*" is the output, v or Δv that timitically results in repetitious input by Pe.

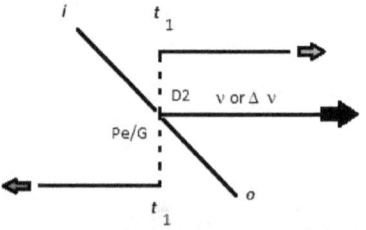

* Potential NOT a function of E, a self-regulating timitic function of the energy's mass transitioning THROUGH space, acting from space as the valve of *df* prerequiring repetitous input to evade the draw of the mass back to equalibrium. Where energy is not transferred to mass manual input.

FIG 1.

Generally, this is due to the oversight that the imaginary reference of a line-segment (also based on energy) represents a section of distance that is 2-directions-in-1. The two-in-one directions are only a result of manual energy transfer; the mass responds with automatic potentiality creating the degree of freedom and thus generally speaking, is the reason behind why the mass is inhibited to 1 *df* as opposed to possessing the ability to access 2 directions simultaneously.

Previously, in my book **Concepts behind Aleodeology** it was my belief that this 2-in-1 distance-time was the reason behind the inhibition of mass to access simultaneity but it was not, it was in fact potentiality of mass automatically established with the establishment of energy through manual input of manual energy.

I do not know if this distance is fractionated (similar to Fractals) without respect to the involvement of energy THROUGH space, since the math also is self-referential, originating from energy based observations. This differentiates itself from Potential, it being the cause of degrees of freedom because manual input of energy to/for a mass or set of masses requires **repetitious input** to evade feedback (Potential/G). Thus the self-referential math derived as well from manually based energy observations, measures the displacement of the input to output THROUGH space, and additionally reveals delimitation to a set of degrees of freedom equidistant to the center of the mass, circulating in paths of 360 degrees that conforms to a sphere. No matter what new area the mass may be relocated to in space, it will always be constrained to this sphere, which it can take up through energy and adopt all rotations within it equidistant from center of the mass or set of masses manually driven by energy, because the *df* are the result of potential in output as well as at equilibrium.

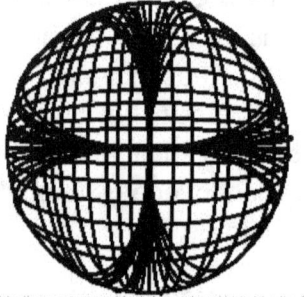

This 'sphere' does not reveal all possible rotations that a mass can occupy throughout the sphere, only a cross-section, here. A rod equi-distant to its center, rotates in 360 degrees, here in 1 direction. Occupancy through all *df* can transpire, but not simultainiously, neither d1 and d2, which are identical.

All rotations would make this diagram appear black, lest colored individually. This sphere represent all *df* that a mass could take reciprocally (360 degrees) aqui-distant to its center, repetition can occur upon compleation of all permutations of a set. The oppositing firection identical to the first can only be accessed by a complete halt, then changing direction to take up all combinations and permutations. The *df* originate from potentiality timitically actuated whenever É is active at the same instant.

FIG 2.

Imagine this as analogous to a die going through all combinations and permutations equidistant to the center of its mass, when is completes a set of permutations there are repeats.

Systems science, contextually, refers specifically to self-regulating systems that are "self-correcting" through <u>feedback</u>; systems science is analogous to space without respect to energy`s pass-through it and would qualify as a self-regulating system or dimensionally interdependent system-(s) to which is analogous to systems science. It can be viewed as space deserving of a meta-language that would describe it and it`s connectivity toward identifying how potential determines the pre-required repetitious input of manual energy that actuates potential from space through energy/mass and exponentially with more power (acceleration). A meta-language not constrained to manually-based input, which not only results in **df** caused by potential/G, as it relates to space through energy transfer to mass or a set of masses, but also necessitates repetitious input to evade potential/G. That being repetitious input is the result of potential and potential causes **df**.

Remember however that the product of potential in manually established energy transferred to mass in 1 direction is of course a self-regulating system, one that differs substantially from manual, repetitious input and is thereby automatic. Therefore to assume that it is the result of energy manually transferred to a mass or set of masses is to incorrectly suggest that potential is just as manual as the manual input of energy to a mass or a set of masses that necessitate repetitious input to evade returning to an equilibrium state.

Potential/G therefore is the origin of **df** from 2-in-1 directional 2D space, represented by an imaginary line-segment representing an actual section of space as it is measured through energy input to mass to travel THROUGH space coupled to manual input of energy to mass a function of the mass with respect to energy. The time factor is distinctly a function of the measurement of mass conveyed THROUGH space in an amount of time that is self-referential because it is manually input, which result in repetitious input to avoid returning to and decelerating back to a state of equilibrium or so-called rest as it would be defined.

V. A Simple Lesson From Scripture

To be convinced that space is anything more or less than a void is to admit to lunacy.

Job 26:7: *He stretcheth out the north over the empty place, and hangeth the earth upon nothing.*

In other words, space is a void, a nothing. To even begin to suggest that it bends due to mass effecting gravity – is absurd.

Scientists are led to believe that Einstein's theory of relativistic absolutes are factual, proof backed in mathematical evidence, logic and tripe embedded in falsehoods contrived by contradictions which distort the truth discredit logic. In essence, it`s wrong-way round supposing that blue is pink, white is black and green is purple. If we are to believe men as opposed to what the truth indicates, scientists will disregard these truths because they fly in the face of Einstein absolute beliefs that show the secret of space, which as we have shown is absent of time, but rather it is energy that self-refers itself as functioning timitically with respect to its passage THROUGH space.

Believing these self-referential absurdities, is to believe that a straight line is not the shortest distance between two points, and that the Earth squashes those who live on the south end of the Earth and strangely affect those who live on the north end of the Earth; all because the mass of the Earth warps space in the presence of so-called gravity.

Space cannot be bent or warped any more than the shortest distance between two points is an arc or a zigzag; because space is a void, and nothing cannot be bent or warped from the establishment of energy, regardless of its base or manipulative abilities. Therefore it is complete garbage to accept these scientists and the false assumption they perpetuate as facts when it cannot, period.

Our answer then does not rest with manual energy THROUGH space or space itself, but rather with automated energy.

Repeats are the product of potentiality, without it repetition could not "happen". This is WHY simultaneity cannot happen; it acts as a safety valve in the face of time.

In the field of Quantum Tunneling (aka the **Casimir effect**) Quanta (in general photons) "borrow" energy from negative energy running a deficit to tunnel its way through a not so lengthy barrier to achieve passage through the barrier popping out the end, or so to speak.

Commonly this relates to what is known as negative energy densities, a weird kind of realm of exotic so-called energy similar to anti-matter. But it's not really anti-anything to say the least. Think of negative energies with the following analogy:

Zero is at least something, but less than zero is less than nothing.

Assuming this (and one should never assume anything) all sorts of faster than light phenomena would be possible using negative energy densities, however isolation of exotic forms of energy from this conjugate, energy is an impossibility. Sad but true, at least according to value-based mathematics. This has led to all sorts of discoveries in science from EPR to quantum teleportation to wormholes to other impossible time travitic theories.

However, according to the scientific community, potential energy is not connected to negative energy densities or exotic energies. As we have stated potential energy is not a form of energy and the mathematics behind these entirely weird bizarre hypothesis are self-referential.

Now in assuming that logic tells us that potentiality has a lower realm below zero, and that is possesses no energy in it since it is less than zero. Yet according to science, it is connected to energy and therefore unisolatable.

If we go back to the imaginary reference of a line-segment we have 2d/D in 1. It would basically indicate that not only does energy have the potentiality to become actuated (actualized or actual from potential) but also includes lots of other phenomena as well. Which isn't uncommon since it has been observed in various causes from people having potential to be better people or the potential of making a great catch.

First, to understand this concept somewhat more, we must disassociate the term energy from potentiality.

For example, let us assume that for every potential outcome of mass over a predetermined travel route or system of routes, that at each position (position is also an imaginary reference) the potential exists for something to occur, only that in our comprehension, that potential corresponds with displacement or transmogrification of energy into motion through the transfer of manual input of energy, which as we know is automatically accompanied by potential pre-requiring repetitious input to produce the output.

It is the potentiality itself that cripples energy from going beyond FTL as it necessitates repetitious input, with more energy more potentiality.

We can estimate here that potentiality with respect to space gives the ability for actualization. But it is not subject to the rule of one particular system, so how input occurs is just as important to how output will be resultant. We say "will be" with respect to the concept of time as a function of energy.

Energy however is not the only "possibility" for actualization into real terms observed through the senses...there exist other potentials, which once accessed properly can be materialized into technologies similar to magic. So our approach must be different to actualize something from potentiality into observable phenomena.

This differs significantly from the potentiality or potential outcomes on probability and permutations, which relies heavily on information theory. Probability also depends on repeats, outcomes may seem "unknowable" but they are limited to a set of possible outcomes –one outcome results out of a set of all possible as Shannon has insinuated in his theory. Adding more "probables" gives more sets and with more sets more possible outcomes with respect to human factors and manual energy put into the system or systems against potentiality.

Each outcome is a single result, as is information representations, for example on a crap table Big 6 and Big 8 are clearly labeled on the layout. 6 and 8 cannot result simultaneously, 6 or 8 however can. The meaning is the decision, but it is unknown which result will transpire even though all numbers in the set will eventually repeat at some point over time with respect to manual energy input.

Like our previous example potentiality are is the cause of degrees of freedom. This potentiality, like a rod representing am imaginary line-segment can go through all combinations and permutations, once gone through repeats occur and they can occur randomly but with our rod the transfer of energy gives the direction, and all possible rotations can be gone through space in 1 direction, forming a sphere.

Unlike probabilities weather random or directed, potentiality dictates their degrees of freedom with respect to energy in opposition to potentiality causing it. Hence it differs from potentiality because the potentiality is below zero, while possibility is +1 adding more energy related observations with respect to energy. So the approach is wrong as it is with manual energy input.

Yet the fact that there is an unknown factor before actualization suggests that potentiality functions on different axioms that can actualize into being. The problem then is the approach, to actualizing what is required and this cannot be through manual input or the result will be inhibited output.

When we differentiate between possibility and potentiality, potentiality separates itself from probability because it is below zero, we can estimate that the unknown factor is the potential for actualization, but what is possible for this actualization to become possible through potentiality will depend on the approach used to actualize what we see fit, and that can be anything.

In other words if we assume that we use an input which is automated as is potentiality, it would then be feasible to actualize an energy (hypothetical) that could abate, supersede potentiality. Or could even manifest completely new forms of phenomena; phenomena that could do things never dreamed of.

All of course originating from potentiality, and why because with potentiality there is an unknown "before" it is actualized. This factors in that given a different approach, such as ***automated input***, as it differs from manual input, the result-(s) would be also be different and the outcome would be different and as such programmable.

Potentiality is the instance of manifesting that which may be possible as it differs from the probable.

The differential corresponds to the input producing the output. For instance, if the input is value-based, then it is **manual** with respect to energy and therewith the output will always relate to **repetitious input** against automatic feedback of (pe/g), further hierarching, out to reciciprocal (not necessarily in 360 degree arcs) repeats constrained to a set of ***df*** also repetitious by the input to output values of manually-based energy against automatic feedback that creates repetitious input producing that repetitious output.

The differential may be further defined by information theory where: "information is a message out of a set of all possible messages". Where the "set of all possible" messages are probable – as in a probabilistic set of letters or values that repeat (like 2 die for example- (i.e. the set is finite)) (chance/gambling is repetitive because the sets repeat at some random point) and again we re-turn back to value-based manual energy.

Exposing that potentiality differs from probability in that that which may be possible is not probable. Hence the impossibility of possibilities which correspond to phenomena exemplified by faster than light, teleportation, etcetera, is not possible in probability because the possibility for such phenomena is not possible using probability especially with sets related to energy input to systems incorporating manual values.

For something such as these, lie in producing output with an input which may be viewed as automatic and or disembodied from energy based on manual values or manual input of energy.

VI. Energy is Manual, Potentiality is Automatic

Potentiality is Automatic.

Potentiality (Pe) is automatic. **The transpiration is established with the establishment of energy (Ke).**

It therefore follows that energy is an automation associated to **MASS**, and mass only. Energy or its automation apart from manual input to mass would be appear as impossible without mass. That is to mean there is no way to separate mass from its partner mass (the phenomena associated with it), isolate it and fiddle around with it to see how it behaves.

This leads to the assertion that space has nothing to do with so-called gravity as it is a nothing, and nothing cannot be affected by something. This should be clear.

So-called gravity **is** the result of potentiality automatically established when energy is put into any system constituting mass. (Fig. 1). It is erroneous to consider anything else outside of this, because it just doesn't add up and blanks will keep appearing as long as scientists refuse to consider this. These "blanks" are indicators of red-flags…that they need reconfiguring.

Instead of concocting false beliefs that time can be traversed with fantastical space warps that can never be acquired, **the focus is aimed at mass and its manipulation** with respect to feedback or redirecting energy in an **automatic fashion** (as an example time is automatic, yet perceptual, which seems to indicate that it is causing space, not merely coupled to it) consistent with behaviors that actually yield superluminal transport. (the term automatic differs in its meaning from mechanization, like how information is automatic and yet the signal from which messages originate have no physical identification as opposed to communication. However the two seem to come one from the other or be interdependent as are energy and potentiality are, and again seem to have their base in potentiality. Hypothetically speak of course) as potentiality is automatic when energy becomes established, for as we know repetitious input and its resultant hierarchies of repetitious behavior is due to potentiality conjoined with the establishment of energy inhibiting mass to achieve super speeds over vast distance.

The more input of energy – the more potentiality **acts** to regulate the acceleration to a velocity, requiring more energy, and with it more input exponentially until a limit is reached where the energy required is beyond that capable of actual viability.

However if mass is manipulated in a way wherein it can loop or be automated in a manner that diminishes potentiality, perhaps through a feedback loop or automated feedback loop-(s), the mass would **behave differently** where that in that when energy is input (put-in) into the system, the system`s response will abate potentiality, and actually acquire the desired goal-(s) of super-speeds and/or even teleportation, hence a pre-programming of you will.

Mass is the thing which inhibits FTL. And it has been explained what is behind this. This is evidenced in the establishment of energy that when the input of energy from and external, manual force, transfers through a body, it`s responsible unto repetitious input. The inhibition is resultant by mass itself in the form of *Pe* **as an automatic response** balancing or becoming greater with the input of more manual energy to the body, **exponentially causing an opposing drag** (so to speak), and **NOT** a result of any unidentifiable gravitational force from so-called space, which is suspected as time. **Gravity is nothing more than this.** In simpler terms; the mass gains **weight** with more input of energy from any manual source. Repeats in the originating force are needed to stabilize or change velocity in order to avoid potentiality as disengagement of the originating force results in a return back to equilibrium of the mass back to a state of non-energy transfer.

VII. Mass Causes Degrees of Freedom through Potentiality

Mass is the reason behind repetition of all physicalities. (That is to mean spherically). Mass is also the reason why 2 directions cannot be used simultaneously, for to take one and appropriate the other requires that the mass come back to an equilibrium position and be re-accelerate to a velocity.

If we attempt to "isolate mass" from energy, such a disembodiment would fail to hold living organisms foreseen as a future representation of people in a spacecraft – no mass plus infinite energy equates to death. Simple logic here; if instead the mass is held in a self-sustaining equilibrium state at zero while energy becomes or exponentiates to infinity (infinity here means endless measurement or never-ending distance (as in Frank J. Tipler`s rapidly rotating cylinder)) such a system would ultimately fail too due to an outer mass also of infinite proportions sustaining the original mass in an attractive state like that of masses on Earth, which would also result more infinite masses holding those two in place exponentially add infintum. This creates another problem even if such a system could work, it`s preposterous nature itself could not benefit in either case, since mass and energy are intimately interdependent. The interdependency is seen in wave motion of a photon; one can no more isolate *v*, *a* and -*v* from *Pe*, *Ke* and momentum.

The only real, conclusive answer then is to assume that the output is the fruit of the input, the meaning that the output for all exemplifications is manually based and therefore would necessitate an infinite series of infinite manual forces backing the original force transferring the force`s energy to push a body composed of - in a single direction accumulating energy unto ever increasing velocities to a sustainable velocities to achieve rapid interstellar travel or any spin-offs.

Since this cannot possibly be done with any reasonable aspect, the solution resides in altering the way in which manual input produces repetitious input that the mass uses in output. Where the manual input itself is the initial force; repetitious input that produces the transfer of repetitious input produces repetitious output from manual input.

In more direct understandable terms manual input (which is repetitious input) is entirely dismissed and substituted with automated input (self-sustaining) input by, to and from the mass itself. Now while this may seem at first absurd, it is not, and can be accomplished via a feedback from the mass itself.

The feedback or feedback loop can only originate from one source of the mass and this source lies in momentum. By feeding momentum, **m**, from the mass back into the system, one obtains the correct formula for a self-sustaining feedback loop which regulates the 4-velocity displacement candidate for control positively and negatively.

Momentum as in a photon is the arc that transpires between the rapid interchange of **Ke** to pe and back again into **Ke**. Arcs, like rotation, are constant changes in direction in a single direction – but in such an arc of a photon they exemplify crests and troughs rather than completing a full repeat back upon itself.

Such "arcs" are connected to acceleration and deceleration, and in a photon two opposing photons do not cancel when passing through one another lest polarized. All that needs to be done here is agitate the system into constant change-(s)-in direction to stimulate the other motion categories, and feed that back into the system absent of a complete repetition as would be in rotation.

This might be similar to an ignition key starting a car, the rest, if properly adjustable, begins the wheel of self-sustainment for the mass and from here can be amplified for that goal of superluminal transport.

Supposing photons have no mass, change-in direction would indicate mass with change-in direction seemingly induced externally with refection at 100%. If mass is 1, It being neutral positionallly) energy is 0 (at rest) (not manually induced via transfer of energy from an external source unto repetitious input against mass vs. equilibrium). Mass is never zero and energy zero transfer. Energy at rest means that the energy transmogrification (establishment) from rest enables mass to change its current position to position, here from zeron to one. But with repetitious input mass acts to become heavy, a dragging behind energy –the potency to come to light speed with more input against itself. (Mass). If mass did not change position with the transfer in energy to the mass – it would still become heavy. "Position" differentiates by energy perpetuating displacement to mass. A special adjustment in energy then would necessitate that it compensate for change-in position of mass by energy to equilibrate mass with energy rather than mass against energy.

*If we assume this compensation for mass by energy anything less than self-administered (pre-programmed); (automated)/(although initiated manually) it cannot function, since potentiality is mass itself – repetitious input would work to oppose accumulation to elite displacements – dragging as it were mass behind it exponentially like a ball to a chain. **The input to output scheme** would then have to be systematically altered to divorce itself from repetitious input via manual inducement unto repetitious input. Omitting repeats in input to output exclude it. If done via automatic positive feedback, the initiation of it, manually begun would have the issue of partiality since feedback loops take a portion of the output and feed it back into the input, and this therefore is a repetition, which conflicts with omitting repetitious input of any sort to produce a satisfactory output absent of mass opposing energy in the opposite direction to the direction of travel by potentiality, which is the mass itself.

Indication has it that **mass** coupled or backed by manually driven energy is the source behind or cause of **all** repetitious displacement, weather this displacement is inductive of repetitious input unto hierarchical repetition forming spheres equidistantly from the axis, mass causes repletion regardless of external input.

Therefore estimating this, the absence of repletion as any level, weather to establish motion or no, would then suggest that removal of repeating is the key to liberating energy from mass even if timitically in the energy spectrum, since energy is the only thing which is timitic.

Since the opposite of repetition is instance, this indicates that the control lies in the very establishment of a repetition when applied energy to transfers to a mass to establish repetitious input to produce the output of what displacement appears to do rather than what is actually happening. This, in effect would mean that during input of energy to a mass (from an external source also having energy to push or pull or both) a repeat in pre-required in replacing energy back into the system to evade the natural equilibrium the mass has to return back to rest—the input itself must be rapid enough to evade or inhibit the tendency to return to equilibrium –to initially disable the opportunity for the mass to return to equilibrium by the input disengaging it – evading the requirement for the energy transfer to input again when the mass acts to naturally return to equilibrium by the first input. The secondary input is required because mass is weighted to return to equilibrium requiring energy to be put in again to continue displacement either by pushing or pulling.

Amends/Conclusion: my conclusions on the topic is that to travel the vast distance of space, mass for a craft and its occupants would have to remain at absolute zero with accumulation to a specified velocity. Space can in no sense be warp-able, since there is nothing there to grasp in the first place – a void. While I have been unable to deduce exactly how to implement this; I have a backburner theory that it is related to a language akin to any previously known. While Edward Snowden has indicated a shadow Government – it is not too far from truth now to suspect that science too has misaligned truth to cover up secondary projects.

VIII. Recent Findings:

Much of my research over the past 24 years has been written out and transferred here. I don`t think it would be wise to publish this information since no-one will actually believe it without a thorough investigation. The systems today are incomplete - many of the sciences try to solve the problem of mass being halted by Pe as something which may be accomplished through a PUSH system devoid of any PULL system. This basically is the solution. No PUSH system will overcome mass and keep the opposition of Pe to Ke at zero in accumulation to a velocity; nor in any unequivalent oscillation. Only a PUSH in 1d along with a PULL system will enable mass to retain an equilibrium state wherewith it can supersede previously unascertainable velocities. The only problem left in this is what exactly would do just this.

I have yet to uncover it, and such disinformation and debunks are out there it is difficult to lay claim to what it might be that can perpetuate this paper for the automation or Aleodelogical transport as I here do coin.

Second, is some information with respect to the Delta-T Antenna; initially I understood that it will take more than probability to overcome the obstacles involved in starship travel and exploring galaxies, to a level akin to fiction or treknology, or even to begin it. The Antenna and some of the related footage I reviewed on the a tour of the Montauk base did in fact reveal that something strange was actually going on there. But with various indications I got an epiphany that suggested that to do these as well as time travel, etc, it would involve conscious interactions – that is through thought amplification and transmittal of those into actual reality. Like at the Montauk Project seat. The concept is phenomenal and it gave me clues to how the aliens might have done it themselves – they too apparently were stumped on space and time travel, with no real solutions, until they understood it had to be through acts of creation. When you understand this idea, you begin to see how by creating it via the thought amplification process – it could in essence could be plausible. The Antenna, perhaps, is just one piece of that total puzzle, but supposedly the aliens built a centralized chair driving thought through the chair, into a thought amplification system transmitting it into reality. That is how they do it did it to achieve the technologies to move about through space and time. Perhaps from here reverse engineering took place and the system was intermingled or used to advance their society. Naturally this would have drawback as well as benefits and would have to be carefully used.

My favoritism in both these discoveries, would be the second one, since after I realized it; it gave me an insight into star trek TOS as I watched it in conjunction with how such technology might be applicable in such a POTENTIAL future – how it all might work. Even though the ship and her crew encountered alien unknowns and through the use of the chair could wipe out an enemy with the flick of the switch, the crew rather dealt with the problems unless the seat was needed in dire cases.

Now the only problem left is to develop a system or axioms which can provide insight into the extrapolation of the necessary information to piece the puzzle together for the goal. Perhaps not necessarily mathematical, nor necessarily linguistic – but perhaps a language which might enable it`s user to **acquire** the information rather than spending millennia on research; like getting the precise information about particles and what exactly they are and getting the right information back. This might be with the bounds of probability and statistics absent of the energy manual input to output scheme analogously, but it may go much deeper or even beyond that, since uncertainty plays such a big role it is culminating.

IX. Still More to Learn

PHASE CONJUGATE RING AND SPHERE
SOLUTION TO INTERGALACTIC TRANSPORT

a) Review of Aleodeology:
b)
I investigated several methodologies and concepts in hopes of bringing about interstellar transport and it`s various spinoffs. Let us here review that concept which will lead into the, afore description of how this can be accomplished after more than years of research: aka automation of mass and or their subsequent spin-offs;

c) *Aleodeology has to do with travel (If one can call it that) between positions by going back to the exact time in which one left, but in a new position or outcome. i.e.; in t_1 (The instant or static instant in time for energy to manifest) for **df_1** *(Degree of freedom 1), **df_2**, etc. Let us imagine a stationary photon, and that we fill up all (**df**) (all outcomes backward and forward equidistantly) with the same photon at P_1 (Position or place 1) at the same time, (or* *before) we unify them all in all possible outcomes at t_1 within the available df for the photon.*

d) The idea here would be that once all outcomes occupy all possible **df** at t_1 at P_1, they could all then be gyrated in multiple **df**. That is to mean reciprocate in 2**df** as opposed to reciprocate in 1**df** *(where a light ring is moved through all possible combinations and permutations of a sphere). Initially this is comparable to simultaneity where causality is violated. *__A 360 degree light ring, rotating through a circumferential distance can only do so in 1d (1 direction) or 1df (1 degree of freedom). The ring itself may be "moved" and occupy in its rotation, all combinations and permutations of a sphere; but only in 1 direction. If the occupancy of the opposing directions are to be utilized, the rotation of the ring must come to a halt and traverse in the opposing direction to the direction it previously rotated through and then it can occupy it`s opposite directions. (or be turned completely around).__ It cannot, in a normal space environment occupy both directions and both combinations and permutations of the ring's rotation and the movement of it through a sphere, which would be simultaneity.

e) This is remedied by a phase conjugate light ring (or field) which travels in the opposing direction and merging the two.

f) The "construct" of the two opposing fields, properly modulated and tuned, should then be free to be able to be moved through or about a sphere equi-distant from the center of its imaginary axis (the center of the ring).

g) The general concept here is either an interior or exterior field for a craft (bubblization), or forming these fields about a mass of say the shape of a person or say a cedar block.

Continuing on we then review: The field can do this because it is the same field (in the now), but can only be done in 1 direction). Displacement from df_1 to another (**df**) now becomes easy because with this new found freedom via **time travel, (**that is to mean, time travel **at the same time** - leaving and arriving **in the same space**) the intervening distance between the two is eliminated, and if I were in it, I could go into any one of them dimensionally.

NOTE: Tipler's infinite cylinder does not do this; it winds through space, goes back THROUGH time to THE past; comes back from THE past to THE future of THE now. It or a field could be assumed to leave and arrive at the same time, however it does not do this, if it did it would leave and arrive in the same space at the same time and that would require time travel of a different sort.

For the reader this may be difficult to envision but just imagine it as taking multiple exposures of an object in all possible degrees of freedom 3 dimensionally (all angles equidistantly). That said, a body would then take-up all possible outcomes at space 1 by continually arriving to t_1 in that space. In other words we are only arriving at the same time we left.

a)
>These are all one space, cross-sectioned at the precise center of the first face (That is a relative view from the front as opposed to moving off to another view to the side) in a Cartesian coordinate system. Now what of 2 two or more beams of equal measurement an equal distance apart?

b) Of

c) df_1 and df_2?

d) *A Degree of Freedom May be defined as an outcome that occurs in an instant of time

$$\frac{t_1}{p_1} \qquad \frac{t_1}{p_2}$$

e)

f) **Separate?**

g) Are they both united?

h) Indeed they are, in the same time (t_1). Once again intervening distance is eliminated because the time is identical, only the outcomes are different.

i) Thus travel between the two is equal to t_1. However we have not considered the same photon beam. If each time photon beam A is sent back to df1 and df2 and an elapsed time t_2 is allowed to occur and is sent back to df_3, df_4, df_5...to t_1, travel would be instant by t_1 no matter how far the distance.

j) Diverse photons can exist simultaneously in multiple df via time travel; sent back to a different df photon A exists in df_1 at t_1, while photon B sent back to t_1, is in df_2. Both exist in t_1 in df_1 and df_2 (not "or"). The df between the two does not therefore exist, the two become one only that the outcomes are different in t_1. As an analogy, this is similar to dice becoming merged for a single result where 5 OR 9 become 5 AND 9.

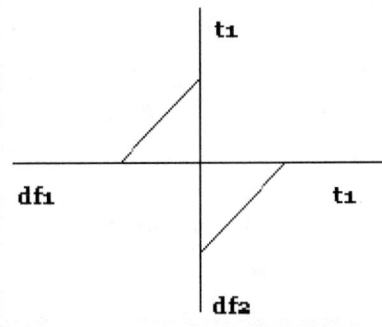

k)

l) **Co-ordinates**

m) This is the *general* concept of Aleodeology.

n) That is, to purpose travel by time travel without having to go very far back *through time*, and arrive at the destination in the past -where light originates - as it takes millions of years for light to travel to reach our atmosphere depending on the distance.

o) In more direct terms, we travel by going back to the same time in a different space, through the utilization of the phase conjugate field (or ring).

p) As odd as this approach may appear it does work and properly implemented can cause other things to move as if directed by magical means; It`s simple and the miniscule requirement for time traversal is only that we move back to the

same time in which we left to move
superluminally or teleportatically.

III. Exhaustive logic combined with detailed writings in my molskine has led me back to the original thought which I had disregarded. This came in the formula that the input being manually driven does cause repetitious output against the opposition of Pe. Both E and Pe are interdependent and mass is in fact the source of gravity; no external mysterious force actually exists. If we assume that the input caused by external source B causes body 1 to displace A, the energy transfer results in the output of motion in the form of v or change in v.

Now –v is the resistance in the form of the masses opposition to itself as it would be with a regulation valve to hinder superluminal transport. Thus source is repetitious to maintain the definition of uniform displacement or change in displacement with more input of power from the source against –v or Pe. Otherwise, however negligible, -v would dominate and the mass would naturally succumb to it upon any disengagement of input from source as well as below a specific power threshold.

If we assume that irrespective of these facts, that the timitic differential between input to output from source which results in repetitious input to evade fallback to –v and non-motion is slight and not instantaneous, then it should be feasible to suggest that in order to overcome this the input necessitates that it first be pre-arranged to be faster than the output to overcome repetitious input from source. This may be interpreted as a source which is not manually established, perhaps self-communicative and automatic. This would indicate that the source itself is repetitious because it too avoids –v fallback. Not a feedback loop, but something different of which I have not yet identified and as of yet am still seeking out. This however is the means through which overcoming the hindrances associated with mass in – v/Pe may be manipulated and overcome unto superluminal travel.

> PHOTOKINETIC CONSTRUCTS (Light Construct)
>
> It is, of course most implausible to think that mass itself can harbor any feasible means through which interstellar transport can reasonably be accomplished. The more you think about it – the more you try, it almost always comes up cold.
>
> However, if we turn our attention to light itself and imagine a thing in the form of light, say for instance a cell phone made entirely from light, we can begin to see that mass is not a solution to space travel, much less teleportation as it would be in fiction. These are what I term as photokinetic constructs or light contructs. To literally build a complete object made out of light which behaves and even simulates a body having the same properties as a mass.
>
> To do so, light itself would have to be able to either self communicate and or time travel or

both. Even if by a few quantum seconds into the past, and then back to the present. This concept has far reaching implications and is far ahead of it`s time –more research is needed

2.

I have contemplated this subject from every angle; from revolving photons,(head to tail not around it) how they might mingle (phase conjugation), enabling stopping through some information source or signaling – but to no avail. My conclusion on this is that intention for the photon to be stationary in space, in order to enable the building of constructs. The only way to do this is only if the photon itself is time travitic. There are two ways; the first is leaving and arriving in the same space and the second in a different space (not too distant from where it left (Aleodelogically). This is the key.

3.

Science has invented probabilities. Potentiality keys into this, but dependency upon it is then one is always defeated just like gambling; if we investigate every available energy form and information, weather it is known or unknown, we come to the conclusion that it IS mass that causes degrees of freedom (df) through it`s own inter-dependency with Potentiality (Pe),which of course is energy. It IS the ONLY reason why gravity is; and NOT any as yet to be identified, everywhere "force". That was a mistake. Remove mass, and this constraint is taken out of the way, but not via obliteration. Potentiality differs from scientific definition; is the solution to FTL, teleportation and many spin-offs.

Mass hinders such goals. Because it relates to energy; and while negative energies "borrow" against E in m, when manually driven, these two borrow to incorporate are the secret so well hidden for so long. Mass therefore CANNOT be involved when potentialities are involved unto FTL and the like; established, or interdependency results.

$$m=df=e=g$$
$$g=Pe$$
$$Pe=m+e$$

This too has drawn upon blanks; perhaps I missed something in my earlier notes seeing that **E** and **Pe** are one and the same, mass. This would require an output which loops 100% Input is applicable – however the output should loop back to alter the input, which would normally be repetitions.

While the above solutions may appear applicable they all come back to flags and blanks. Negative energy, cannot do it because both **Pe** and **E** are interdependent and inseparable. The other potential methods are impossible aspirations. The reasonable way is via automation; that is to mean the source itself must be automatic and not dependent upon manual input, resulting in repetitious output. This could in fact be a language or self-communicating energy, perhaps pre-programmed – it is as of yet not uncovered. Magnetic s can be said to be automatic, as would be other processes such as those seen is sci-fi when Kirk beams down, the transport hints at automation. This is not like automation as it would be with cars – no, this is very, very different.

*Some typographical errors may be evident in this work– please notify the author at the above email and contribute to advance this work.

www.ingramcontent.com/pod-product-compliance
Lightning Source LLC
Chambersburg PA
CBHW070405190526
45169CB00003B/1120